REPAIR OR REPLACE

Clogged drains, leaky faucets, running toilets. These are the all-too-frequent plumbing problems every homeowner faces. They seem too minor to go to the expense and inconvenience of paying and waiting for a plumber, yet you are not sure where to start. Just follow the steps in this quick guide, and you'll discover how easy these repairs are to make yourself.

And often enough you look at that leaky old faucet and think, wouldn't it be nice to upgrade? Maybe you are craving one of those new kitchen faucets with a sprayer that pulls conveniently out of the spout. Quite likely, you'd pay a plumber more than the price of the new faucet to do the installation. This job does involve some squirming around under the sink, but as you'll see in the steps provided here, there is nothing complicated about it. The tools are simple—a screwdriver or two, an adjustable wrench, maybe a basin wrench. Today's PVC drainpipes and flexible supply tubes streamline the job. Many new faucets attach to the supply lines with quick connects that are literally a snap to install.

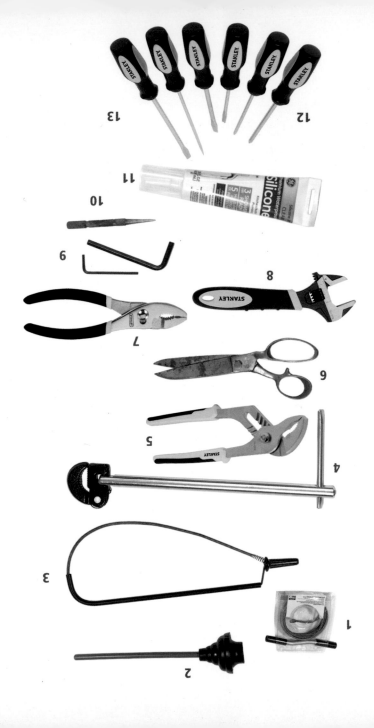

WHAT YOU'LL NEED

- Drainpipe auger for unclogging a sink [1]
- Flanged plunger for unclogging a toilet [2]
- Toilet auger for unclogging a toilet [3]
- Basin wrench for hard-to-reach nuts between the sink and the wall (usually needed only for replacing faucets on deep kitchen sinks) [4]
- Adjustable pliers for loosening tight hand nuts and installing a ball cartridge [5]
- Scissors for cutting the collar off a new flush valve and cutting refill tube to length [6]
- Slip-joint pliers with two positions for grasping and turning [7]
- Small adjustable wrench for water-supply connections and other nuts [8]
- Allen wrenches for removing the handles from single-handled faucets and replacing the seat for brass cartridges [9]
- Nail set for fishing out faucet valve seals and springs [10]
- Silicone for under gaskets if the top of the sink is pitted [11]
- Phillips-head screwdriver for most of the screws you'll encounter [12]
- Thin-blade flat-head screwdriver for lifting the decorative caps on faucet handles [13]

CLEARING A CLOGGED SINK

You can try using a plunger or liquid drain cleaner to clear a sink clog. If those approaches fail, you have two other options: You can disassemble the P-trap under the sink, or you can clear the clog with an auger. With modern PVC plumbing, disassembling the P-trap is often the easiest and most effective option.

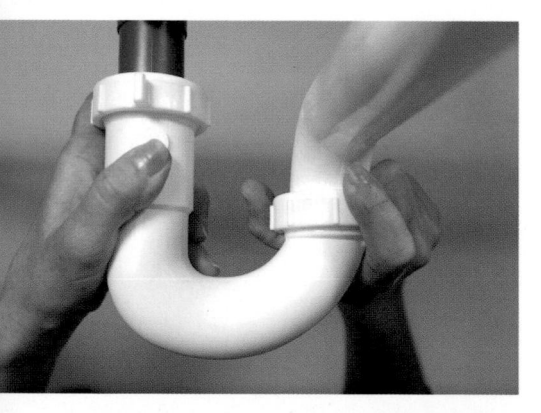

DISASSEMBLE THE P-TRAP.
One connector secures the trap to the trap line going into the wall. Another connector fastens the drain tube from the sink. Simply unscrew the connectors by hand, clear the blockage, and reassemble.

CLEARING A SINK
WITH A DRAIN AUGER

1. REMOVE THE POP-UP LEVER. If you are working on a bathroom sink with a rod that operates the stopper, you'll need to go under the sink to remove the pop-up lever. Unscrew the cap, pull out the lever, and then go topside to remove the stopper.

2. CLEAR THE DRAIN. Insert the boring head of the auger into the drain until you hit the clog. Slide the handle to a point near the sink and tighten the set-screw. Rotate the auger to clear the clog.

CLEARING A CLOGGED TOILET

You have two options to clear a clogged toilet: using a plunger or a toilet auger.

USE THE RIGHT PLUNGER. If you are lucky, one good plunge will clear that clogged toilet. The secret to success is to use a plunger with a flange on the bottom like the one shown here. The flange makes a much better seal on a toilet.

USE A TOILET AUGER. These augers are designed to navigate the serpentine trap built into toilets. Pull the cable up through the plastic sleeve, and insert the sleeve into the toilet. Then turn the handle to rotate the cable as you work it through the waterway. When you hit the obstruction, keep rotating to break it up.

STANLEY

REPAIRING A LEAKY SINGLE-HANDLE FAUCET

Single-handle faucets use either a ball or a cartridge. There are two types of cartridges. Older faucets, including many single-handle tub faucets, have longer cartridges, often with replaceable rubber rings around the cartridge body. Newer cartridges typically are shorter and have durable ceramic disks inside.

BALL OR CARTRIDGE? From left to right, a ball, an old-style cartridge with rubber rings, and a newer cartridge with ceramic disks.

The procedure for removing a ball or a cartridge is the same—just follow the first three steps on the facing page. Then see either "Repairing a Ball-Type Faucet" on pp. 8–9 or "Repairing a Cartridge-Type Faucet" on p. 10.

1. TURN OFF THE WATER SUPPLY. With the faucet valves on, turn the water off at the hot- and cold-water-supply valves under the sink.

2. REMOVE THE HANDLE. Pull out the little decorative cap, and use an Allen wrench to back off the setscrew until you can pull the handle off.

3. REMOVE THE COLLAR. Just grip the collar and unscrew it by hand.

> **QUICK TIP** Don't remove the setscrew from the handle. On some faucet handles, it can be fiddly to get back in place. Besides, if you leave it in place, you won't lose it.

4. PULL OUT THE BALL OR CARTRIDGE. If there is a ball, you'll find a plastic cap with a rubber gasket fitted to it. Pull this off and remove the ball. Some older-style cartridges have a retaining clip that you'll need to pull out with pliers. Otherwise, just pull out the cartridge by hand.

REPAIRING A BALL-TYPE FAUCET

1. REMOVE THE SEATS AND SPRINGS. Once the ball is removed, use a nail set or similar tool to remove the two pairs of seats and springs. Replace them and any other parts that seem worn, especially any rubber parts that have become brittle.

2. REASSEMBLE THE SEATS AND BALL. Install the new springs and seats, then insert the ball, aligning its slot with the pin in the faucet body.

3. PUT THE CAP IN PLACE. With its gasket in place, align the tab on the cap with the notch in the faucet body.

4. SCREW ON THE COLLAR. The ball may need to be pressed down against the springs so the cap tab can engage the notch to prevent the cap from rotating when you install the collar. With the tab aligned, put the collar in place. Use pliers to press down the cap while you start screwing on the collar.

REPAIRING A CARTRIDGE-TYPE FAUCET

1. REPLACE GASKETS.

A ceramic cartridge has three holes on the bottom. Sometimes there are separate O-rings for each hole, or there might be a gasket that goes around all the holes. Replace what you find—the cartridge itself rarely wears out. Replace the rubber rings on the body of an older-style cartridge, or if it is worn, replace the cartridge itself.

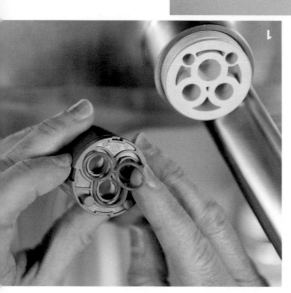

2. REASSEMBLE THE FAUCET. Put the cartridge back in place, and reinstall the clip if there is one. Screw on and hand-tighten the collar. Put the handle in place, tighten the setscrew, and pop the setscrew cap back into place.

REPLACING A SINGLE-HANDLED FAUCET

Single-handled faucets are the most popular type these days, especially for kitchens. There are a huge variety of styles to choose from, but they are all installed in pretty much the same way. Installation details vary, however, so be sure to read the instructions that came with your faucet.

When choosing a faucet, make sure it will work with your sink's hole configuration. Most kitchen sinks have three holes for the faucet, but some have only one faucet hole.

The instructions that follow are for a kitchen sink. If you are installing a single-handled bathroom faucet, make sure the holes in your sink are spaced to accept a "centerset" faucet as discussed on p. 22. Also, for a bathroom faucet, you'll need to hook up the stopper as described in "Installing a Two-Handled Faucet."

QUICK TIP If you are replacing the sink along with the faucet, it's a lot easier to attach the faucet to the sink before installing the sink. Also, you can simply detach the old faucet from the water supply and dispose of the old sink and faucet as a unit.

1. DISCONNECT THE WATER SUPPLY.
Turn off the hot- and cold-water-supply valves. Open the faucet valves to drain. If you will be reusing the water-supply tubes, use a basin wrench to disconnect them from the faucet stems.

▶

Do You Need the Old Supply Tubes?

■ In most cases, one end of each supply tube is attached to a water-supply valve, and the other end to a stem coming from the faucet. The connections to the valves are usually more accessible and easier to disconnect with an adjustable wrench as shown. Some new faucets, including the one being installed here, are made with supply tubes that come out of the bottom of the faucet hub and go directly to the valves, eliminating the need for separate supply tubes. With these, disconnecting at the supply valves is the obvious choice. Flexible supply lines made of braided stainless steel or plastic are inexpensive, so you might decide to replace the existing tubes in any case, especially if you see any corrosion at the connection.

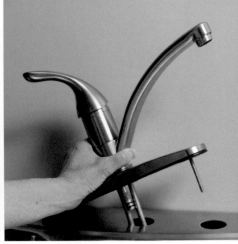

2. REMOVE THE OLD SPRAYER. You might need a basin wrench to unscrew the sprayer hose from the faucet hub. The one shown here has a quick connect. Just squeeze the tabs and pull down. Remove the locknut that attaches the sprayer to its hole. Pull out the sprayer.

3. REMOVE THE OLD FAUCET. The faucet is secured with two plastic locknuts. You should be able to loosen them by hand, but you might need to grab them with pliers. Remove these fasteners and pull out the faucet.

▷

STANLEY

4. INSERT THE THREE-HOLE ADAPTER. The faucet shown can be installed in a one-hole sink and comes with an adapter for installation in a three-hole sink. Put the gasket under the adapter, and secure the adapter underneath with locknuts.

5. PUT THE ESCUTCHEON AND HUB IN PLACE. Put the escutcheon in place over the adapter. Insert the rubber gasket under the hub, slip the hoses through the hole, and align the hub with the tabs on the adapter.

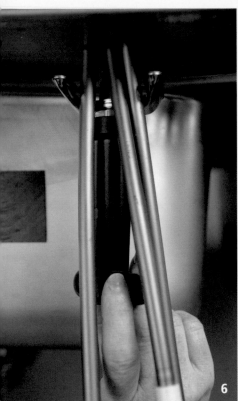

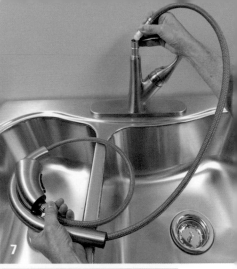

6. SECURE THE FAUCET TO THE SINK. Under the sink, hand-tighten the locknuts. This faucet comes with a U-bracket. Slip the U-bracket onto the hub, capturing the three tubes as shown. Then use a wrench to tighten the center nut. This faucet came with a plastic wrench, shown here, for the center nut. An adjustable wrench or basin wrench will work, too.

7. ASSEMBLE THE SPRAYER HEAD AND SPOUT. Attach the sprayer head to the threaded end of the supplied hose. If supplied, put the little tapered cap called a hose guide on the other end, and feed it through the spout and hub. Slide the spout onto the hub. ▷

8. INSTALL THE HOSE.

A donut-shaped weight hangs on the sprayer hose to help the hose retract into the faucet. Slide the weight over the hose, then push the hose end onto the short tube coming out of the middle of the hub. Install the clip over the connection as shown.

9. CONNECT TO THE WATER SUPPLY.

If the tubes are too long, you can put a loop in them, but the loop must have a diameter of at least 8 in. Connect the tubes to their respective hot- and cold-water-supply valves and hand-tighten. Then use a wrench to tighten one more turn.

QUICK TIP

If the faucet you are replacing had a separate sprayer, you can use the extra hole to install a soap dispenser. Some faucets come with a dispenser.

10. FLUSH THE LINES AND CHECK FOR LEAKS. Open the shut-off valves. Remove the sprayer head, and pull the hose into the sink. Turn the faucet handle to the mixed position, then flush the water lines for one minute. This will prevent any debris from getting into the sprayer head. Reinstall the sprayer head, and tighten the threaded connection as shown. Tighten any connections that leak.

REPAIRING A LEAKY TWO-HANDLED FAUCET

All two-handle faucets use some kind of cartridge to regulate water flow. If the handle screws down, you have an old-style brass cartridge that presses a rubber seal into a replaceable brass seat. On newer faucets, you'll find a plastic cartridge that slides over a rubber seat with a spring beneath it.

OLD VS. NEW. Old-style cartridges (left) are brass; newer cartridges are plastic.

The sequence for disassembling your faucet will be essentially the same as shown here, although trim pieces can vary a bit. The cartridges do vary in the rubber O-rings and seats. When you pull the cartridge, just leave all that stuff in place and take your cartridge to the hardware store to find the right repair kit. Then replace every part that comes with the kit and reassemble the faucet.

Find the Problem

To determine which faucet valve is leaking, turn off the cold-water-supply valve under the sink. If the leak stops, the problem is the cold-water faucet. If the leak doesn't stop, turn the cold supply back on and turn off the hot. If the leak stops, the problem is the hot-water valve. If not, both valves need repair.

QUICK TIP Before you start, make sure the sink is stopped or put a towel over the drain just in case you drop any small parts.

1. REMOVE THE CAP. Turn off the water at the supply valves under the sink as shown on p. 7, step 1, and open the faucet valves. Use a thin-blade screwdriver to pry the decorative cap off the faucet handle.

2. REMOVE THE HANDLE. Remove the screw that attaches the handle to the cartridge and take off the handle.

STANLEY

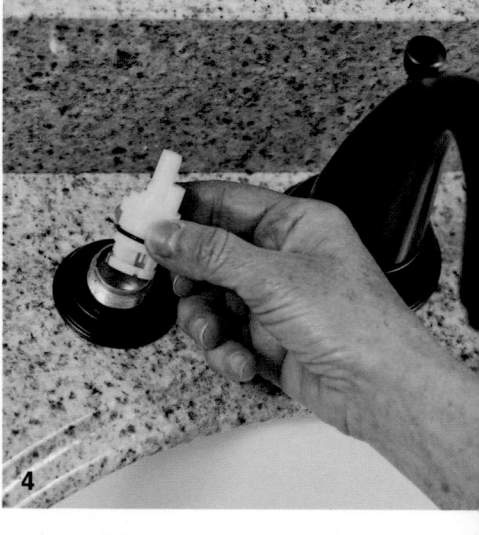

3 **4**

3. REMOVE THE RETAINING NUT.
Use an adjustable wrench to
remove the nut that holds the cartridge down.

4. PULL OUT THE CARTRIDGE. This
usually requires no tools. You
might have to grab the cartridge
with pliers and jiggle it a bit if it
is stuck.

5. REMOVE THE SEAL AND SPRING. For a plastic cartridge, use a nail set or similar tool to pull out the spring and seal. For a brass cartridge, use an Allen wrench to remove the brass seat.

6. REASSEMBLE THE FAUCET. Replace the parts that came with your repair kit, then simply reverse the steps above. Make sure the raised area at the top of a plastic cartridge is facing in toward the spout.

INSTALLING A TWO-HANDLED FAUCET

The two-handled faucet installed here fits in a bathroom sink with three holes spaced in a configuration called "widespread," meaning the handles are separate from the spout and 8 in. on center from each other. If the handle holes are spaced 4 in. on center, you can use either a "centerset" faucet, in which the handles and spout are one unit, or a "mini wide-spread" faucet with separate handles. Although installation is essentially the same for all two-handled faucets, details vary, so be sure to read the directions that came with your faucet.

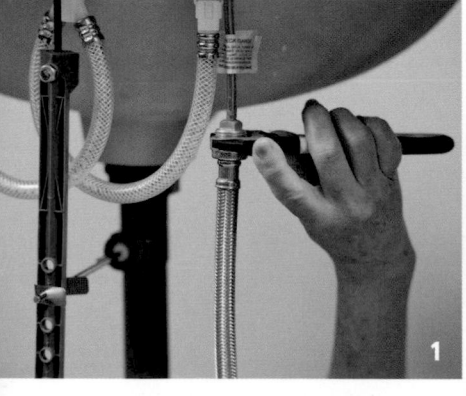

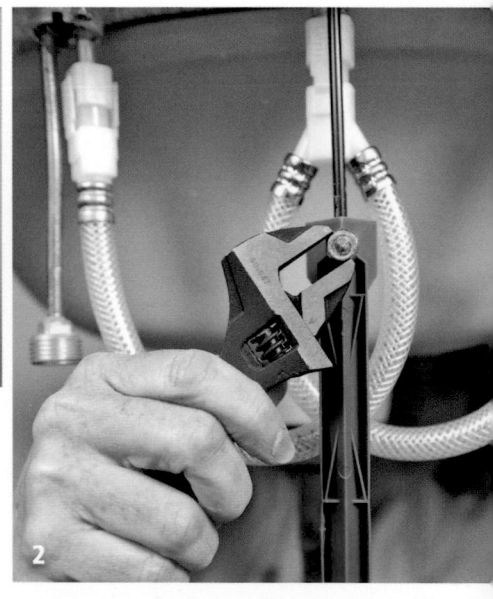

1. DISCONNECT THE WATER SUPPLY.
Turn off the hot- and cold-water-supply valves. Use a wrench to disconnect the supply tubes from the valve stem. See pp. 11–12 for other ways to do this. Open the faucet valve to release pressure, and make sure the stopper is closed to prevent small parts from dropping down the drain.

2. REMOVE THE STOPPER ROD.
Loosen the setscrew that secures the stopper rod to the strap. Then pull the rod out of the back of the faucet.

3. DISCONNECT THE MIXER TUBES. Double-handled faucets have tubes that bring the hot and cold water from the faucet valves to a junction fitting that sends water through the spout. Disconnect this tube from the faucet valves and spout using a wrench, if the connection fittings have nuts. This one has quick connects. Just pull out the tabs to release the connections.

4. UNFASTEN THE VALVES. If there are set-screws securing the locknuts under the sink, loosen them. Then remove the locknuts and washers. Pull the valves out of the sink.

▷

5. REMOVE THE SPOUT. Use an adjustable wrench or basin wrench to remove the nut securing the spout to the sink. Pull out the spout.

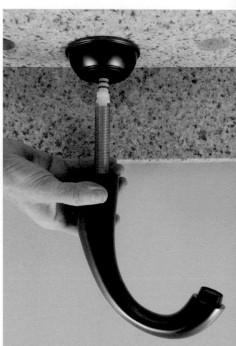

6. INSTALL THE NEW VALVES. Make sure the area around the sink holes is clean. If the sink top is uneven, put a bead of silicone on the gaskets under the faucet valves and the spout base. Put the valves and spout in place. Under the sink, secure the faucet valves with the washers and locknuts provided. If the nut has setscrews, tighten them.

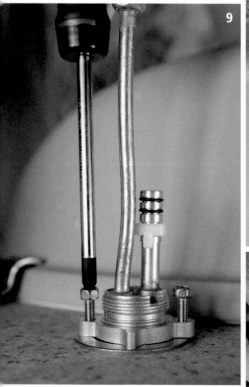

7. INSTALL THE SPOUT AND SUPPLY.

Insert the spout in its hole, and slip the stopper rod through the back of the spout. Under the sink, put the washer and nut in place. The washer is slotted to fit around the stopper rod. Secure with an adjustable wrench or basin wrench. Don't overtighten—you could crack the sink. Connect the water-supply tubes to the faucet valves, and tighten with a wrench.

8. INSTALL THE MIXER HOSE AND ATTACH THE STOPPER ROD.

Snap the mixer hose connector onto the spout. Cross the hoses to make the bends gradual, and snap the connectors into the faucet valves. Slip the stopper rod into the strap, and tighten the setscrew.

QUICK TIP While you are tightening nuts under the sink, have a helper up top make sure the faucet valves and spout are centered over their holes and the handles are lined up correctly. If nobody is around, get everything snug and check topside before tightening.

STANLEY

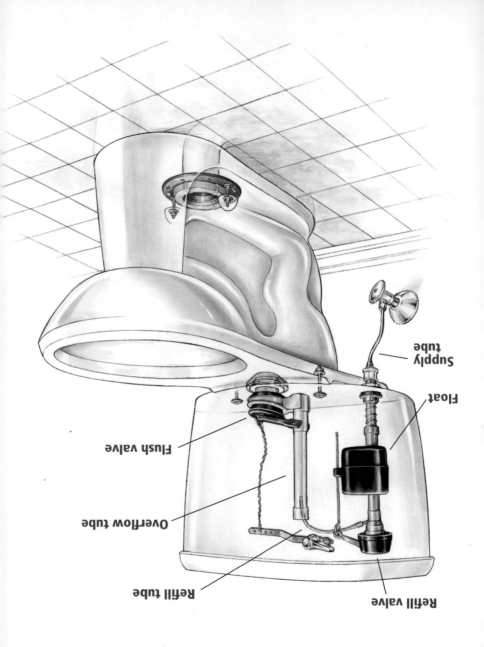

Supply tube

Float

Flush valve

Overflow tube

Refill valve

Refill tube

FIXING A RUNNING TOILET

There are a number of different flushing mechanisms that might be in your toilet tank, but they all work the same way. When you push the flush handle, it pulls up a flush valve, also called a flapper, and allows water to rush into the bowl.

When you release the handle, the flush valve closes. Water comes in through the refill valve, filling the tank while a refill tube pours water into the bowl through an overflow tube. As the water flows into the tank, it raises a float of some kind. In the two most common configurations, the float either travels up the valve body as shown in the drawing on the facing page, or the float is a ball on an arm. In any case, when the float reaches a certain level, it causes a lever to shut off the fill valve.

If the toilet is running continuously, it is usually because water is pouring into the overflow tube because the float is set too high. See "Adjusting the Float" on p. 28. If the water is trickling into the bowl, perhaps causing the toilet to flush on its own, it's likely because the flush valve isn't sealing. See "Cleaning or Replacing the Flush Valve" on p. 29.

Sometimes the problem is that the entire fill valve and float mechanism is encrusted with minerals that are preventing parts from moving. It's cheap and easy to replace the whole thing, as described in "Replacing the Fill Valve and Float" on p. 30.

ADJUSTING THE FLOAT

1. ADJUSTING A BALL FLOAT. The ball float shown has two adjustments. For large adjustments, you can loosen a plastic wing nut to adjust the angle of the ball arm. Most likely turning the screw at top counterclockwise will be enough. Some older balls don't have adjustment screws. These have metal arms that you can bend down to adjust the ball.

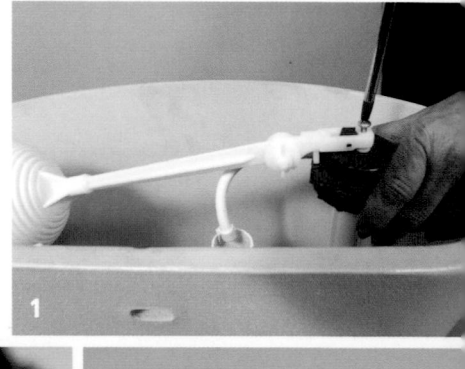

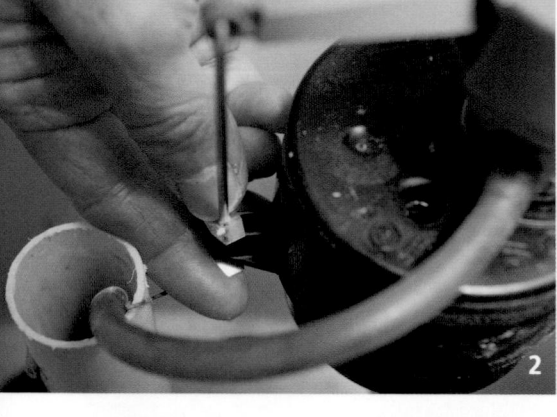

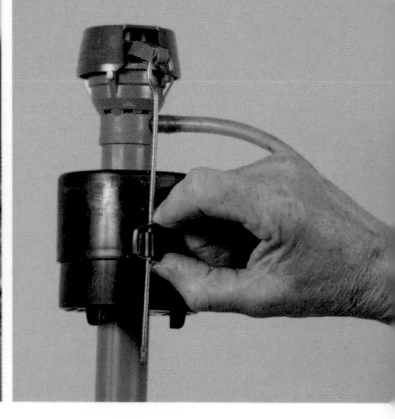

QUICK TIP Newer toilets are designed to use less water than older models. If you want to save water with your older toilet, try moving the adjustment clip down an inch or two, adjusting until the flush is just powerful enough to do the job.

2. ADJUSTING A VALVE BODY FLOAT. If you have a float that travels up the valve body, the float is likely secured to a metal rod that's connected to the refill valve lever. Squeeze the clip and move the float down 1 in. When the tank fills, the water level should be about 1 in. from the top of the refill tube. If it's not, repeat the procedure, adjusting the float until you get it right. Newer mechanisms often have a knurled knob on top or a different type of clip for adjusting the float height.

CLEANING OR REPLACING THE FLUSH VALVE

1. REMOVE THE FLUSH VALVE. Turn off the water supply to the toilet, and flush to empty the tank. Pull the flush valve off the pegs on the overflow pipe. Instead of attaching to pegs, some flush valves have a collar that fits over the pipe.

2. CLEAN OR REPLACE THE VALVE. Try using a sponge to clean the flapper and its seat. If the valve still leaks, disconnect its chain or strap from the flush handle and replace the valve. If your overflow pipe has pegs, use scissors to remove the collar from the new flush valve. Put it in place and attach the chain or strap to the handle.

REPLACING THE FILL VALVE AND FLOAT

1. DRAIN THE TOILET.
Turn off the water supply to the tank, then flush the toilet to drain it. Use an old towel or rags to soak up as much of the remaining water in the tank as you can.

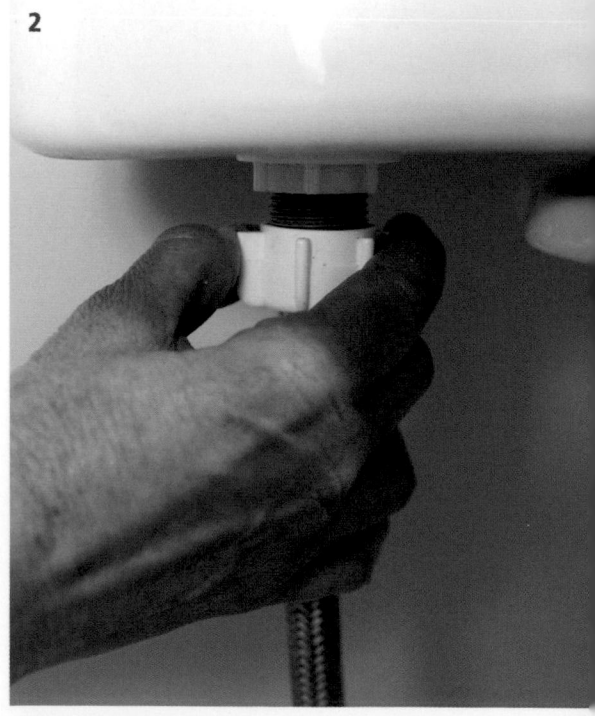

2. DISCONNECT THE SUPPLY.
Disconnect the supply tube from under the toilet. Use pliers if the nut is too tight to loosen by hand. Remove the refill tube from the overflow pipe. Under the toilet, use pliers to remove the nut holding the refill valve in place, then pull the valve out of the toilet along with the ball if there is one.

3 **4**

3. INSERT THE NEW REFILL VALVE.
With the bottom seal in place,
insert the new refill valve into
its hole.

4. SECURE THE NEW VALVE. Under
the toilet, hand-tighten the nut
on the valve. Reconnect the
water supply.

▷

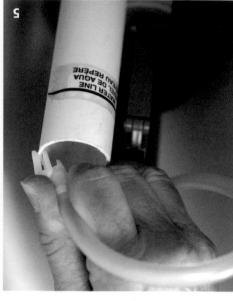

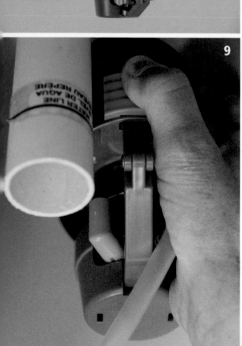

5. ATTACH THE REFILL TUBE. Slip the new refill tube onto the valve. If necessary, cut the refill tube to length. Clip or insert the tube into the overflow pipe. Turn on the water, flush, and make sure there are no leaks under the toilet. If there are, loosen the nuts securing the valve under the toilet, reset the valve, and retighten the nuts.

6. ADJUST THE FLOAT. When you flush, the water should shut off when the level reaches about 1 in. below the top of the over-flow tube. If necessary, flush the toilet to drain and adjust the float height. On this particular model, you depress a button on the side of the float to move it up and down (rather than using an adjustable clip).